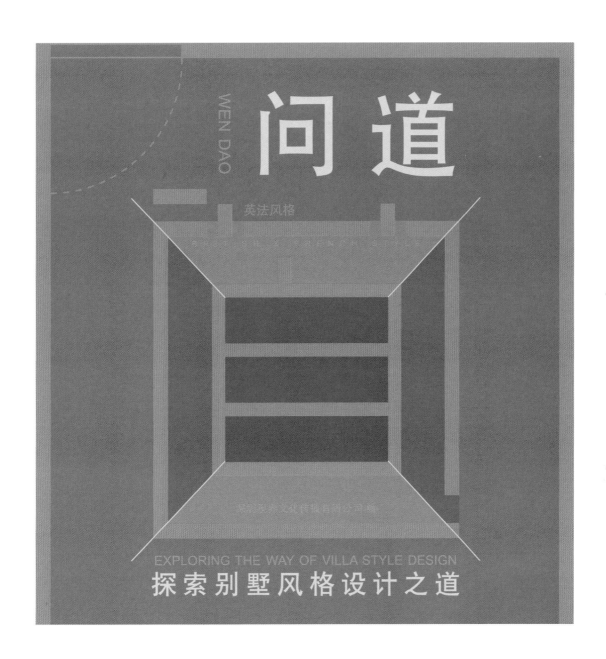

问道
WEN DAO

英法风格

EXPLORING THE WAY OF VILLA STYLE DESIGN

探索别墅风格设计之道

中国林业出版社
China Forestry Publishing House

图书在版编目（CIP）数据

问道：探索别墅风格设计之道：英法风格 / 深圳视界文化传播有限公司编． -- 北京：中国林业出版社，2017.1
　　ISBN 978-7-5038-8918-9

Ⅰ．①问… Ⅱ．①深… Ⅲ．①别墅－室内装饰设计 Ⅳ．① TU241.1

中国版本图书馆CIP数据核字（2017）第023432号

编委会成员名单
策划制作：深圳视界文化传播有限公司（www.dvip-sz.com）
总　策　划：万绍东
编　　　辑：杨珍琼
装帧设计：潘如清
联系电话：0755-82834960

中国林业出版社　·　建筑分社
策　　划：纪　亮
责任编辑：纪　亮　王思源

出版：中国林业出版社
（100009 北京西城区德内大街刘海胡同 7 号）
http://lycb.forestry.gov.cn/
电话：（010）8314 3518
发行：中国林业出版社
印刷：深圳市雅仕达印务有限公司
版次：2017 年 2 月第 1 版
印次：2017 年 2 月第 1 次
开本：215mm×275mm，1/16
印张：18
字数：300 千字
定价：280.00 元（USD 48.00）

PREFACE 序言

CREATING THE TIME OF VILLA STYLE DESIGN
打造别墅风格设计的时代

With the development of modern China's economy, villa has been welcomed by more and more people in some big cities. The development of modern villa has been changing with each passing day. Villa is not only a good place for vacation, but also a kind of spiritual culture which can cultivate people's tastes and improve people's living standard. On function, organization and content, villa has been changing constantly. Personality design, ecological design and the use of high-tech have become the new trend of present villa design. And in the whole villa design, the interior design style determines the overall pattern of the villa.

Style is the tone and manner that presented by an age, a genre or a person's literary works in ideological contents and artistic forms. In the work by Roman writers Terence and Cicero, it refers to "type of book" and "type of writing", meaning some certain ways in which words express thoughts. The "style" in English and the "stil" in French and German come from here. However, the "style" in the series of books Exploring the Way of Villa Style Design has nothing to do with pure arts such as painting and sculpture, instead it pays more attention to artistic features, creative personalities and era characteristics about home decoration designs. In other words, what it shows is the "style" of daily life and a display of artistry of living space.

The style discussed in this book is the most noble and romantic British and French styles which pay attentions to interspersing architecture in nature, which is luxurious without piles. The details present exquisiteness and the luxury manifests elegance. One can enjoy the romantic pastoral life and pursue the relaxing comfort and cozy. The design emphasizes on natural belongings of souls, which gives people a fresh sense. The open space structure, ubiquitous flowers and green plants and exquisite furniture create a noble, romantic and comfortable living atmosphere of the overall design.

Here through the perfect displays with pictures and words, we can intuitively enjoy the historical features and cultural characteristics of British and French styles. They are no longer abstract concepts conveyed by words, but living, visible, cultural and material existences. Follow up, let's go to enjoy the romantic encounter of the noble prince and the elegant princess.

随着现代中国经济的发展，在一些大城市别墅越来越受到大众的欢迎，现代别墅的发展也是日新月异。别墅不仅仅是度假的好去处，更作为一种精神文化，从而陶冶人们的情操提高人的生活水平，在功能上、组织上、内容上别墅不断发生着变化。个性设计，生态设计，高科技的使用成为当前别墅设计的新趋势。而在整个别墅设计里，室内风格设计决定了别墅的整体形态。

风格，即一个时代、一个流派或一个人的文艺作品在思想内容和艺术形式方面所彰显出的格调和气派。在罗马作家特伦斯和西塞罗的著作中，该词有"书体"、"文体"之意，指文字表达思想的某种特定方式。英语、法语中的"style"和德语中的"stil"皆由此而来。不过，《问道——探索别墅风格设计之道》这一系列丛书所探讨的"风格"，与纯粹艺术如绘画、雕塑之类无关，而是关注家居装饰设计的艺术特色、创作个性以及时代特点。换句话说，它所呈现的，是日常生活中的"风格"，是对居住空间的艺术展示。

本书所探讨的是风格之中最为高贵、浪漫的英法风格。讲究将建筑点缀在自然中，奢华却并不堆砌，细节中体现精致，华丽中尽显典雅。享受浪漫的田园牧歌式生活，追寻放松的舒适与安逸。在设计上讲求心灵的自然回归感，给人一种扑面而来的浓郁气息。开放式的空间结构、随处可见的花卉和绿色植物、雕刻精细的家具等，所有的一切从整体上营造出一种高贵、浪漫、舒适的居家氛围。

在这里，通过图文并茂的完美展现，我们可以直观的欣赏到英法风格的历史特性与文化特征，它们将不再是文字所传达的抽象概念，而是活生生的、肉眼可见的、洋溢着文化气息的物质性的存在。随着编者的脚步，让我们一起去细细品味这一场高贵王子与优雅公主的浪漫邂逅。

CONTENTS 目录

BRITISH & FRENCH STYLE 英法风格

008　FRENCH NOBILITY
　　　法式贵族

020　THE GOLD LAND
　　　金色大地

032　A SNOWY KINGDOM
　　　雪色王国

046　FLOWERS BLOOMING IN A DREAM OF BLUE
　　　梦之蓝 花尽开

054　FAIRYLAND—PARIS AT MIDNIGHT
　　　天境——浓情午夜巴黎

072　INTERPRETING NEW ROMANCE IN FRENCH STYLE
　　　经典演绎法式新浪漫

090　MEETING FASHION CLASSICS OF PALACE
　　　邂逅于宫廷般的时尚经典

106　A PSALM
　　　如诗的礼赞

132　NOBLE STYLE IN BRITISH MANSION
　　　贵族格调 英式豪宅

150　REAPPEAR THE NOSTALGIC STYLE AND ELEGANCE OF DOWNTOWN ABBEY
　　　再现"唐顿庄园"的复古与优雅

166　THE RHYME OF WHITE POND LILY
　　　白莲花的诗韵

190　DEEP CHANT OF BLUE
　　　海蓝深咏

198　DEDUCING ARISTOCRATIC PALACE TEMPERAMENT
　　　演绎宫廷式贵族气质

208　CULTURED AND REFINED HOME
　　　博雅之轩

216　A CLOUDY MANSION
　　　云·邸

230　DIGNIFIED ENJOYMENT OF LIFE
　　　尊贵生活享受

248　THE BLUE DREAM IN THE MANSION
　　　庄园蓝梦

260　THE SEA OF PARIS
　　　巴黎的海

272　A MONUMENTAL MANSION
　　　传世大宅

BRITISH & FRENCH STYLE

英法风格

Exquisite and romantic
natural and comfortable

Dignified and elegant
luxurious and noble

精致浪漫　　自然舒适

端庄典雅　　豪华贵气

008 BRITISH & FRENCH STYLE 英法风格

DESIGN COMPANY | 设计公司
北京意地筑作装饰设计有限公司

LOCATION | 项目地点
北京

MAIN MATERIALS | 主要材料
维纳斯灰大理石、世纪米黄、壁纸、木地板、木饰面等

DESIGNER | 设计师
连志明、张伟、徐辉

AREA | 项目面积
280m²

FRENCH NOBILITY

法式贵族

DESIGN CONCEPT | 设计理念

The designers blend French neo-classical style into simple and practical modern design to make it become an elegant neo-classical mansion with better functions and beautiful sceneries, which is suitable for modern people to live. Both the furniture and accessories depict the elegant and noble status of the owner with graceful and beautiful posture as well as gentle flavor full of connotation. The common fireplaces, crystal palace lanterns and Ionic columns are the highlights, while the elegant and noble atmosphere is pervaded in the concise design, which contains fashionable elements in every detail, creating an ideal residence for elites to nurture the body and soul.

设计师将法式新古典主义风格中注入简洁实用的现代设计，让其成为适合现代人居住，功能性强并且风景优美的新古典主义美宅。无论是家具还是配饰均以其优雅、唯美的姿态，平和而富有内涵的气韵，描绘出居室主人高雅、贵族之身份。设计师用常见的壁炉、水晶宫灯、爱奥尼柱作为点睛之笔，简练的设计中渗透着典雅的贵族气息，点滴细节中蕴涵着时尚元素，缔造出一个精英人士修养身心的理想居所。

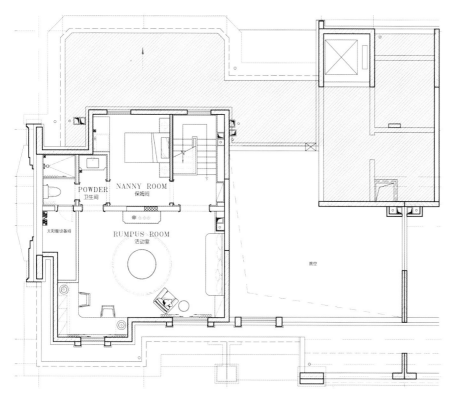

SPACE PLANNING | 空间规划

入口玄关处的墙面使用壁纸与法式的玄关柜相结合，艺术挂画、插花自然的融入空间。设计师将原建筑楼板拆除，将整个空间做成挑空。客厅与餐厅，中西厨区域相贯通，形成一个大的开敞空间。客厅区域通顶的书柜加壁炉的组合，两个爱奥尼柱子，更加凸显整个区域的空间感。卧室空间布局紧凑合理，独立的衣帽间、卫生间、配套齐全，体现出主人高雅贵气的生活方式。

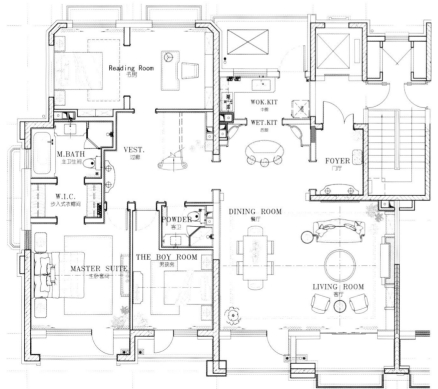

客厅：

　　墙面以清爽优雅的象牙白护墙板与传统纹样的米色壁纸相结合。整个空间以柔和的米色、蓝色、金色为基调，低调而高雅贵气。同时挂画上融入文艺复兴时期的复古元素，增添了空间的时代气息。

餐厅：

　　米色与蓝色的搭配，再加上巧妙的融入自然安静的山水风景挂画，意境悠长。餐桌上温馨的花束配以少许枯枝，造型别致，与挂画的自然气息相呼应。

主卧：

　　整个色调以米色打底，蓝色丝质面料点缀，传统的法式地毯与挂画相呼应，使整个空间散发着浓厚的高贵气质。细腻雅致的床头背景墙面分为三个部分，以中间对称，磨砂镶以金色线条的纹理镜面，大气奢华，衬托着主人的尊贵。

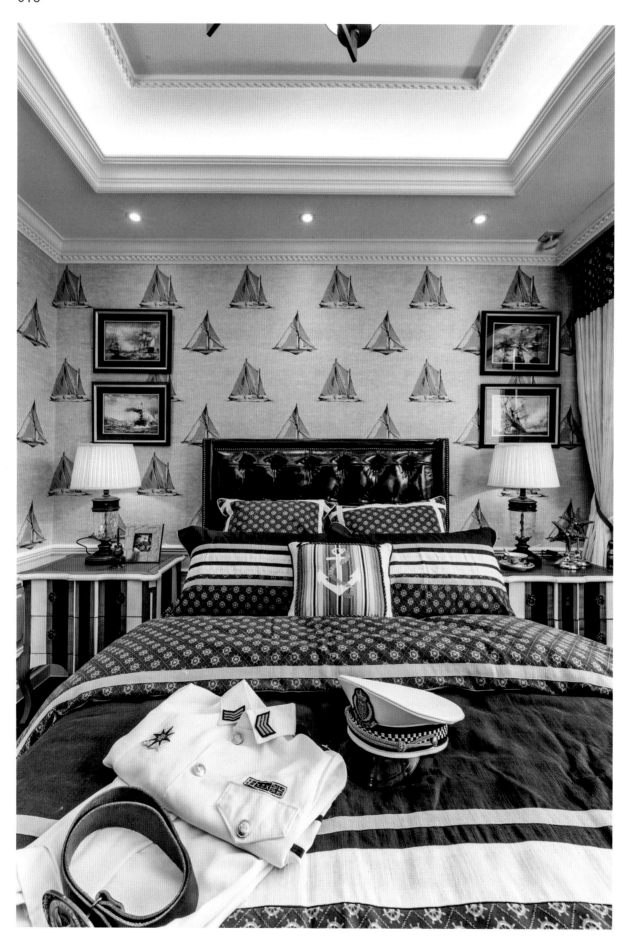

男孩房：

　　整个色调以海军蓝为主，房顶部的白色与蓝色相呼应，凸显了主题，清爽自然。海军风格的床品与挂画，是房间的独特亮点，搭配上浴室的装饰摆件，令人仿佛能呼吸到大海的气味，男孩子的活泼和勇敢也在对海的崇拜与向往中自然流露。阁楼作为孩子的活动室，延续了男孩房的军事主题，在如此一个小天地里，孩子定能尽情展现自己的艺术天赋。

DESIGN COMPANY | 设计公司
筑土都市设计咨询有限公司

LOCATION | 项目地点
北京

MAIN MATERIALS | 主要材料
大理石、壁纸、玻璃、地毯等

DESIGNER | 设计师
王雯

AREA | 项目面积
400m²

THE GOLD LAND
金色大地

DESIGN CONCEPT | 设计理念

This is a two-floor duplex mansion with an area of 400 square meters, and the area of the ground floor and the underground floor that has been counted into the sales area are the same. Combined with the architectural construction and hard decoration style of this project, Archiland Consultant International hopes to awaken the senses with colors and furnishings to build this project into a residence with three generations in French neo-classical style, creating a luxurious manner of a mansion. In the glorious space, you will not feel overcautious. No matter details or the whole, the design of the overall space achieves a visual balance, overflowing an atmosphere which is luxurious yet not proud, enthusiastic yet fresh and refined.

这是一套面积为400平方米的两层复式公寓大宅，地上地下各一层，面积相同，且地下一层计入销售面积。结合此案的建筑构造和硬装风格，筑土设计希望用色彩和陈设来唤醒感官，将此案打造为法式新古典的三代居，营造出富足的大宅风范。在这华贵的空间中，你不会感到拘谨。整个空间的设计无论是细节还是整体都做到了视觉的平衡，洋溢着富贵不凌傲、热情却脱俗的气息。

SPACE PLANNING | 空间规划

设计师从居住者的整体需求出发，分别配置了客厅、餐厅、厨房、卧室等主要活动区域，同时贴心考虑到家庭成员的年纪，依据他们的不同喜好，量身打造了各自的舒适区。

客厅：

在客厅中，设计师使用名贵的布料、丝绒质感的靠包、晶莹的水晶吊灯，以及充满法式风情的装饰品，传递出一种温馨的奢华。家私的选择上，是以中式及法式风格结合现代风格来搭配。中式元素，在这个整体上偏法式的空间中，占据了很重要的一部分。它们以挂画、花鸟图案、摆件、家私等不同的形式呈现出来，也凭借强烈的色彩视觉感进行统一，让这个法式大宅呈现出不一样的中式情结。

玄关：

　　整个画面延中轴线几乎呈严格对称分布，以大地色为主的色调更具大方庄重之美。而金色镶边的红色挂画的加入则打破了原有的持重感，让原本属于过渡空间的玄关有了截然不同的魅力。

COLOR MATCHING | 色彩搭配

设计师运用了大量大地色系，令所有空间融合起来。又在不同的区域搭配宝蓝、深紫、暗红、橙黄等色彩进行点缀，每一个空间都透出不同的韵味和感受。作为空间里重要部分的卧室，设计师也采用了不同的色调。老人房中，为了迎合年长者的心理需求采用了金秋色，营造出安静宜人的养心环境。男孩房则采用明亮的蓝和橙，希望积极向上的色彩辅助男孩子的心理成长。主卧中浅紫色与暗红色的组合，让色彩表现得更加优雅，床幔的装饰增加了贵族的气息，加上极具艺术魅力的刺绣床品，烘托出了空间的雍容华贵之美。

主卧：

　　主卧房色调沉着，层次丰富，中式纹样的靠枕图案与陶瓷插画摆件相得益彰。一抹白色绒毯的加入，提亮了整个空间色彩，为主卧房添加柔美细腻的情调。

032　BRITISH & FRENCH STYLE　英法风格

DESIGN COMPANY | 设计公司
重庆碧涛装饰设计公司

PHOTOGRAPHER | 摄影师
麒文空间摄影 张骑麟

AREA | 项目面积
350m²

MAIN MATERIALS | 主要材料
大理石、壁纸、布艺等

A SNOWY KINGDOM

雪色王国

DESIGN CONCEPT | 设计理念

This show flat is dominated with romantic French style, which follows French romanticism. The overall design technique is elegant, noble and romantic, while the style tends to be dignified and generous. The space design stresses symmetric layout, which is steady and magnificent with noble and elegant space perception. On details, carvings and lines are adopted with delicate consideration. French columns, carvings and lines are used in the space, whose craftsmanship is mature and perfect. Entering the door, you will feel a romantic and fresh sense. The elegant and gorgeous tone forms a light and romantic home with French flavor.

"This is one of the best architecture, where we are standing while we don't know where is the end of nature and the start of art." Actually the interior decoration is the same. The transformation of this space is just like a snow in early winter which makes the world white and bright within an eyewink. Then the hearts calm down. The interior decoration continues French style. The glided and carved French sofa and graceful lines interpret the gentle beauty of the space with beautiful modeling and concise crafts. The designer creates an interior environment full of romantic life and artistic atmosphere through delicate and mellow soft decorations.

　　本样板房采用了浪漫法式的设计风格，沿袭法国浪漫主义色彩，整个设计手法优雅、高贵、浪漫，风格则偏于庄重大方。在空间设计上讲究平面布局的对称，稳重大方的气势，高贵典雅的空间感受；在细节处理上采用了雕花、线条，每处工艺都经过精细的考究。空间运用了法式廊柱、雕花、线条，制作工艺完美娴熟；推门而入，浪漫清新之感扑面而来，淡雅、亮丽的基调形成轻盈浪漫的法式情怀之家。

　　"最好的建筑是这样的，我们深处其中，却不知道自然在哪里终了，艺术在哪里开始。"其实室内装饰也是同样的理解，这个空间的幻化就像一场初冬的雪，只是一眨眼，这个世界变白了变亮了，心里忽的就安静下来了。室内装饰还是延续法式格调，镀金雕花的法式沙发，优美的曲线诠释了空间的柔美，造型唯美，工艺简洁。设计师通过细腻、柔美的软装搭配营造出充满浪漫生活情趣和艺术氛围的室内环境。

客厅：

　　客厅多用明快清新的白色与蓝色，中间缀以小色块的棕色，配置简约的线条，既有古典的豪华与典雅，又更适应现代生活的休闲与舒适。大型水晶吊灯莹莹生辉，是设计师简单里透出尊贵、典雅中浸透豪华的设计哲学表现，也传达出居住者追求品质、舒适生活的人生态度。

门厅：

门厅以洗练纯净的白、高贵优雅的灰为主色调，灰白交错，配以温暖的米色，平衡出无限清新舒适。流畅的空间线条、自然简单的纹理花样，塑造典雅、大气的空间。设计师运用空间本身的特性，于方寸间融入大格局的思考。

钢琴室：

白金配的基础用色尽显庄重气息，而蓝色窗帘搭配则为空间注入清新活力。设计师将考究的搭配延伸到空间的每个角落，黑色光面钢琴、白色雕像、金色墙面装饰、旁逸斜出的花束等细节，形成丰富而极具变幻的感官效果。

主卧：

　　主卧内所有陈设与软装配饰格调相同，在颜色选择上，以蓝色为基础色调，各种深浅浓淡的蓝充盈了整个空间，在米色和灰色背景墙的烘托下，空间的素净淡雅氛围更为突出。一束黄色花卉的点缀，跳跃出空间的温暖生动。在图案上则选择自然气息浓厚的花色和藤蔓样式，与整体空间氛围更为搭配。

COLOR MATCHING ｜ 色彩搭配

色彩的搭配上，设计师采用了白色与米色的主基调，结合优雅的浅蓝色、清新的浅绿色、高贵的浅灰色作为空间内的点睛色，营造出一个浪漫巴黎风情的居室空间。结合造型雅致的线条，做工精致的家具，整个空间呈现出一幅恬淡、浪漫的法式画卷。设计师选用明媚、光鲜的色彩装饰来搭配出悠闲、安逸、清闲的慢生活节奏。

SPACE PLANNING ｜ 空间规划

客厅开放的空间中没有多余色彩的堆砌，白色和蓝色是空间的主色调。在主卧区，设计师充分利用已有空间大小，整合规划出便于屋主休憩的放松空间。空间与空间衔接自然，连贯流畅。

046 BRITISH & FRENCH STYLE 英法风格

PROJECT NAME | 项目名称
中海上海紫御豪庭复式样板房法式浪漫

LOCATION | 项目地点
上海

AREA | 项目面积
480m²

DESIGN COMPANY | 设计公司
上海李孙建筑设计咨询有限公司

DESIGNERS | 设计师
孙明亮、陈小玲

MAIN MATERIALS | 主要材料
黑色擦旧亚光漆、金箔漆、樱桃木高光漆、古铜金等

PHOTOGRAPHER | 摄影师
陈盛

FLOWERS BLOOMING IN A DREAM OF BLUE

梦之蓝 花尽开

DESIGN CONCEPT | 设计理念

Located in the central core area, The Amethyst covers inner Hongqiao area, Gubei New Area and belongs to Changfeng section, it's a rarely high end project which integrates ecology, business and entertainment within the city centre. As the first residential project in Changfeng ecological business district, The Amethyst is a masterpiece after 33 years' brand test of China Overseas Estate and 20 years' effort of Shanghai branch, striving to create a scarcely villa and high-quality exquisite duplex. French romance originates from art styles of France in the 18th century. Carved patterns, delicate lines, finely crafted furniture, exquisite fabrics, ubiquitous floriculture, as well as some romantic elements create a beautiful picture of leisurely French life. The layout highlights on the symmetry of axis, making a magnificent momentum and a luxuriously comfortable living space.

中海紫御豪庭地处中环内核心区域，辐射内虹桥、大古北，位于长风板块，是市中心内不可多得的集生态、商务、娱乐于一体的高端项目。中海紫御豪庭作为长风生态商务区首个纯住宅项目，是集中海品牌33年千锤百炼，中海上海公司20年积淀之力作，着力打造稀缺性别墅和高品质精装大复式。法式浪漫起源于法国18世纪的艺术样式。雕花、精致的线条、精雕细琢的家具、精美的布艺、随处可见的花艺以及一些浪漫的元素营造出法国人对待生活悠然自得的美好画面。布局上突出轴线的对称，营造出恢宏的气势，打造豪华舒适的居住空间。

客厅：

　　法式浪漫清新情调是本案给人的第一视觉观感。以白色为基调，加入优雅的水蓝色和富贵的金色。结合大自然的青绿色植物，力求表现精致、悠闲、舒畅、自然、浪漫的生活情趣。加上家具的陈设与布置，创造出自然内敛的氛围。大面积的翠蓝色墙面、水蓝色的绒布沙发、白色的碎花抱枕，让人瞬间产生了自然的回归感，给人一种扑面而来的浓郁清新气息。而金色镶边工艺，恰如其分，含蓄地表达了主人对高品质生活的热爱。造型逼真的水晶烛台吊灯，金色的底座、白色的灯光配以水晶串珠盘旋其间，浪漫而富有情调。而地面洁白的瓷砖搭配花色优雅的地毯，让空间因为绚丽的色彩搭配，色泽清丽且层次丰富。

SPACE PLANNING | 空间规划

　　充分利用空间的通透感,各个功能区域的转换合理贴切,既可以扩大视觉范围,又能很好的利用自然景色。每个部分都精心勾勒,为屋主提供一个既舒心又唯美的享受空间。

餐厅：

  整体布艺以水蓝色调为主，风格以纤巧、精美、浮华、繁琐为主，讲究与自然的和谐。餐厅与客厅均以绒布沙发为主，配合精致的碎花工艺，柔美精致。大面积落地窗在花白相间的透明纱窗映衬下，自然光线交融，室内外景致互相交流，充分享受美食与美景。富有质感的纯水蓝色布艺窗帘，柔软细滑，搭配盆栽的绿色植物，使空间充满生机和层次感。

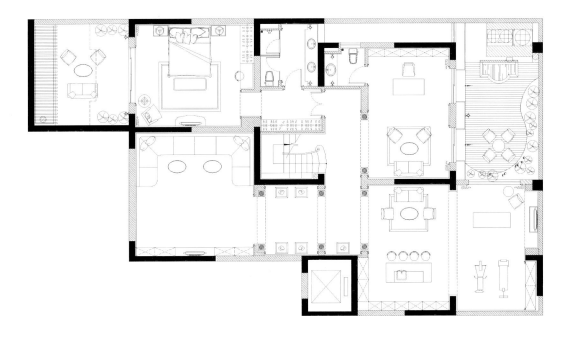

卧室：

以布艺创造温馨与舒适的私密空间，让梦乡徜徉在浪漫的田园之中。多层次的浅蓝色窗帘及白色窗纱搭配，让进入房间的阳光也过滤出生机。线条流畅色彩清新的浅蓝色布艺床品，精致的金色大木床，让大自然与人更加贴近。讲究的是自然回归的感觉，追求一种安逸、别致、舒适的生活氛围。

054 BRITISH & FRENCH STYLE 英法风格

PROJECT NAME | 项目名称
中梁国宾一号太洋房别墅
LOCATION | 项目地点
浙江温州
AREA | 项目面积
640m²

DESIGN COMPANY | 设计公司
上海西麦装饰设计工程有限公司
CHIEF DESIGNER | 主案设计
许蓓蓓
MAIN MATERIALS | 主要材料
丝、陶瓷、铜、水晶、天鹅绒、桃花心木、金箔等

GUIDE DESIGNER | 指导设计
林赛赛
PHOTOGRAPHER | 摄影师
陈雪峰

FAIRYLAND—PARIS AT MIDNIGHT

天境——浓情午夜巴黎

DESIGN CONCEPT | 设计理念

This is the most luxurious estate project created by Zhongliang Real Estate Group in Wenzhou. The luxurious and comfortable villas customized for successful people have once created sales miracles. The orientation of this project is the successful man's gift for his beloved wife, themed as romantic classical French style. The main tone is white and beige brown with fresh light blue, graceful light purple and pleasant light gray as the highlighted colors in the space, which creates a romantic residence with Paris flavor. The graceful lines and decorations, such as walnut, mahogany, basswood and ebony, are dominated with decorative techniques of carving, gilding, inlaying of wood, porcelain and metal, and so on.

  本案系中梁集团在温州重金打造的极为奢华的地产项目，为成功人士量身定制的奢适别墅空间，曾一度创造了销售奇迹。它定位于成功的先生向宠爱的太太献礼，以浪漫的法式古典风格为主题，甜蜜幸福的氛围始终洋溢其中。白色与米咖色的主基调结合清新的浅蓝色、优雅的浅紫色、怡人的浅灰色作为空间内的点睛色，营造出一个浪漫巴黎风情的居室空间。优雅的线条和装饰，胡桃木、桃花心木、椴木和乌木等，以雕刻、镀金、嵌木、镶嵌陶瓷及金属等装饰方法为主。

SPACE PLANNING ｜ 空间规划

设计师一切以女主人的需求为重点，除了必备的生活空间外，还布置了一个完全属于她自己的私人会所，让女主人能在此休闲、聚会。在工艺上设计师除了运用精致细腻、反复雕琢的细节，以及鎏金饰彩的装饰与摆件，还将空间羽化成富有感情的人，传递着深深地爱恋。

客厅：

装饰雅致的高级铜配瓷摆件、施华洛世奇水晶吊灯等奢华材质配合精细雕刻的家具来诠释经典浪漫的法式空间。空间处理上，吸取了洛可可艺术的不少特征，并带有浓郁的贵族宫廷色彩，精工细作，富含艺术气息。豪华独特的水晶灯饰以及清新脱俗的窗饰软搭更加强化了主体，利用线光源和点光源的变化，丰富了空间感觉，完善了设计平面的层次。整个空间，让人感受到真正的人文关怀和无以伦比的奢华高贵。这些奢华温柔的元素静静融合，整个空间呈现出一幅恬淡、浪漫、高贵的法式画卷。

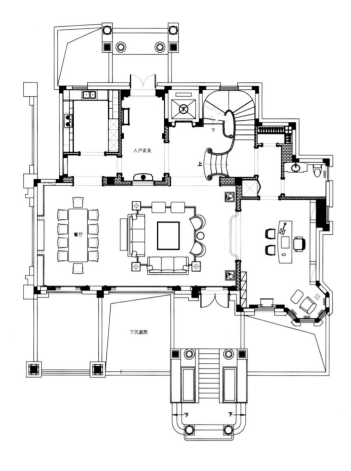

书房：

　　天花吊顶在这个空间中有了更丰富的表现形式，无暇白色加上花朵造型的装饰线条将空间柔美细腻的女性气质展露无遗，同时赋予空间强烈的立体感。配合晶莹剔透的水晶吊灯，奠定整个空间精致浪漫的空间氛围。棕色拼花座椅、黑白棕交错的马赛克地毯、深褐色木质地板，设计师运用不同材质串联相似的色彩，循序渐变，深浅有度，组成一幅和谐融洽的书房图。

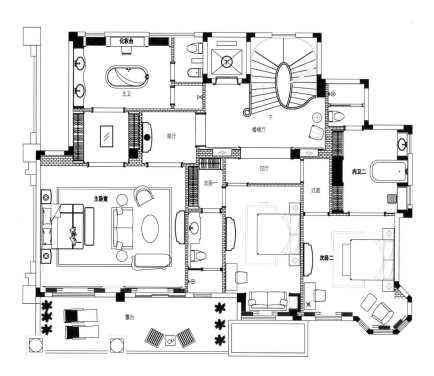

主卧：

空间藉由两块矩形天花板与长条隔断屏的造型，强化空间与线条的理性关系，极具层次感。空间大而开阔，家私摆设简而精，巧妙极致的心思、考究的细节，演绎着空间独属的奢华与时尚。

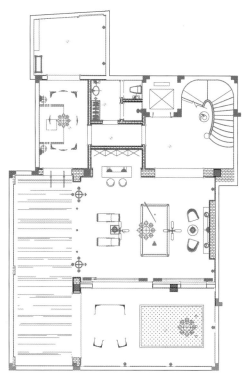

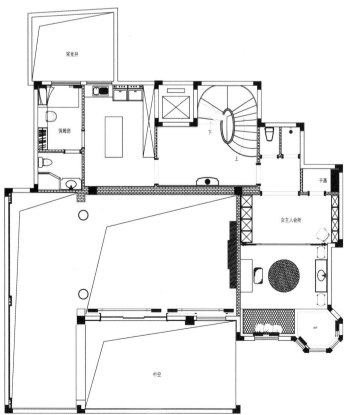

PROJECT NAME │ 项目名称 淮安红豆美墅	**DESIGN COMPANY** │ 设计公司 上海益善堂装饰设计有限公司	**HARD DECORATION DESIGN** │ 硬装设计 汤玉柱
LOCATION │ 项目地点 江苏淮安	**SOFT DECORATION DESIGN** │ 软装设计 宋莹	
AREA │ 项目面积 400m²	**MAIN MATERIALS** │ 主要材料 仿古砖、木饰面、进口墙纸、大理石等	

INTERPRETING NEW ROMANCE IN FRENCH STYLE

经典演绎法式新浪漫

DESIGN CONCEPT │ 设计理念

The orientation of delicate French style endows the space with more expectations. Symmetry is stressed on the overall layout, which is comfortable and full of noble temperament. Elegant French carvings and lines are used in the details, inheriting the essence of French style with refined crafts. Bright off-white color is dominated on the wall, while the furniture is based on dark colors such as red, blue and brown, collocated with perfect curves, embodying a unique and elegant luxurious life. The colors are saturated with full layers, giving dizzying surprises to the inhabitants. A number of large French windows give the biggest visual impression of the single villa. Dense and thick branches add a natural sense into the original luxurious life.

精致的法式风格定位，赋予了空间更多的期待。整体布局上突出对称,豪华舒适又饱含贵族气质。细节处理上运用了精美的法式雕花、线条，制作工艺精细考究，传承了法式风格的精髓。墙面都以明亮的奶白色为主，家具则选用深色系的红色、蓝色以及褐色来凸显，配合完美的弧线曲度，展现了一种别致雅趣的奢华生活。整体色彩饱和，且层次丰富，给居住者应接不暇的惊喜。而多个大面积落地窗的设计，给了独栋院落最大的视觉发挥。层层叠叠、枝桠妖娆，给原本华美的生活添了一份自然气息。

DESIGN CONCEPT ｜ 设计理念

本案是一座独栋别墅，满庭绿植环绕，四季风景明丽动人。穿过庭院，别墅设置两个入门口，加强室内与室外空间的交流。一层客厅、餐厅、厨房相连，富有质感的金色和蓝色搭配更显示出空间的品质感，餐厅的露台设计加大室内采光，并能在就餐时享受屋外美景。二层分布着家庭室、儿童房和客房，休闲区域的设计适合品茗、下棋和闲谈，打造享受型的慢生活。三层则是男女主人的专属区域，卧室连带着卫生间和衣帽间，同时还给主人设计了书桌和化妆台，主卧套房规划合理、精细，满足主人对生活的所有的期待。

客厅：

　　映入眼帘的蓝色给人视觉上的冲击，深灰色的沙发和蓝色、红色沙发单椅，色彩跳动，却又和谐统一，壁炉采用车边镜装饰，周围大理石有着法式风格的雕刻，巴洛克式风格镜子与墙面镜拓展空间视野，天花上的椭圆形雕花演绎出法式风格的精魂。

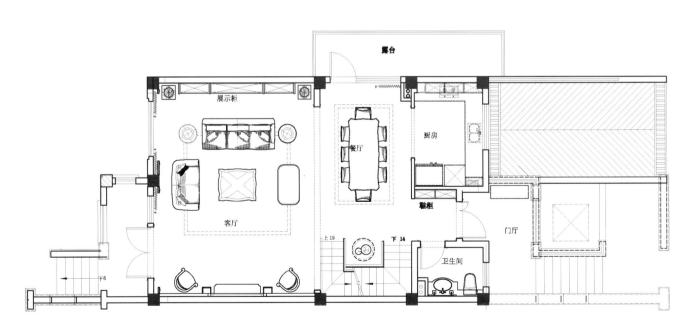

餐厅：

  餐厅与客厅相连，于是便延续了客厅的配色和风格，保持了一层空间调性的一致。金色与蓝色搭配的餐桌椅，餐桌上的玫瑰则营造了空间浪漫的进餐氛围，戴帽式的吊灯有着水晶装饰，水晶在灯光的照射下更加熠熠生辉。

卧室：

卧室呈现出华丽的美感，金色和黑色拼接的花纹床品，床头板有着细美的雕花，淡色带花的壁纸清雅，竖条蓝白相间的墙面易延伸空间视觉，金色斗柜上的相框和插花又赋予了空间生活气息，阳台上的休闲椅也是闲暇时的好去处。

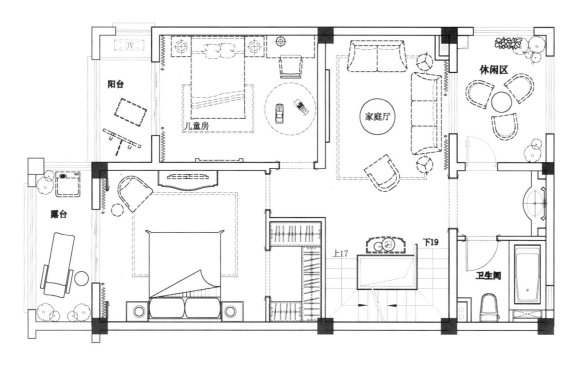

男孩房：

　　橙红色和深蓝色的搭配让男孩房更显年轻和动感，墙面装饰富有线条感，采用米色突出蓝色和橙色的靓丽。吉他、篮球抱枕和棒球帽，突出男孩张扬的青春与活力，水晶吊灯则保留了法式的精美和典雅。

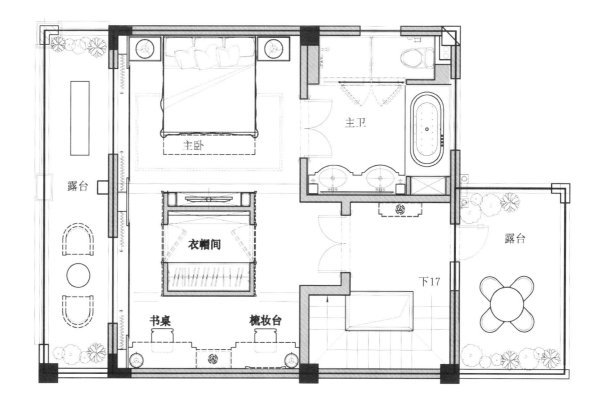

BRITISH & FRENCH STYLE 英法风格

PROJECT NAME | 项目名称
品时空

LOCATION | 项目地点
上海

MAIN MATERIALS | 主要材料
大理石、金箔、手工雕刻、实木薄片镶嵌、布艺等

DESIGN COMPANY | 设计公司
上海全筑建筑装饰设计有限公司

AREA | 项目面积
363m²

MEETING FASHION CLASSICS OF PALACE

邂逅于宫廷般的时尚经典

DESIGN CONCEPT | 设计理念

The project uses the design of thick household Italian Renaissance style, which flows luxurious, dynamic and changeable visual effects everywhere. Meanwhile, it absorbs aesthetic details from the rococo style, uses lively colors and fine decorations which are delicate and tend to be complicated, and shows off a different life taste and a palace life interest everywhere. Fashion and classics encounter in Italian Renaissance, showing the noble and elegant design philosophy in the depth, which becomes a reflection of the owner's enjoyable life. Based on the principle of three crafts, namely manual sculpture, solid wood chips set and pasting gold foil, the designer creates a top household environment, which is classic, fashionable, extravagant and comfortable. The owner can just enjoy the honorable privilege of palace!

　　本案以浓厚的意大利文艺复兴风格的家居为设计，处处流动着豪华、动感和多变的视觉效果。同时也吸取了洛可可风格中的唯美细节，采用明快的色彩和纤巧的装饰，精致而趋于繁琐，处处炫耀着与众不同的生活品味和宫廷生活趣味。时尚与经典邂逅于意大利的文艺复兴，于深沉里显露着尊贵典雅的设计哲学，成为屋主享乐生活的一种写照。以手工雕刻、实木薄片镶嵌和贴金箔三大工艺为原则，打造经典、时尚、奢华、舒适为一体的顶级家居环境，享受宫廷般的尊贵礼遇！

客厅：

　　客厅有着古典精神的气质，体量大的沙发厚实大气，以金色镶边处理。红木色的墙体镶嵌，圆形的吊灯采用金色灯臂，使古典中透露着些许奢华。客厅中的烛台和艺术摆件，以黑色和古铜色为主，增添了客厅的艺术气息。

餐厅：

餐厅采用灰黑色地砖斜铺而成，增强视觉效果，吊灯富有特色，古朴中透出悠远的历史感，墙面巴洛克式的装饰镜有着精美的雕饰，天花上的射灯设计，有意突出镜子的艺术美感，木色的餐桌温润富有质感，能促进就餐氛围。

SPACE PLANNING | 空间规划

　　从宽敞的大厅进入室内，迎面的走廊在吊灯的照射下为回家增添了仪式感。一层分布着客厅、厨房、餐厅、收藏室、家庭活动室和一间卧室，收藏室三面设计了嵌入式玻璃墙柜，柜中陈列着珍贵的瓷器艺术品，在黑色大理石和墙柜的衬托下，同时在柜中加入射灯，更加凸显了收藏品的精美。二楼以卧室休闲区域为主，卧室统一都采用黑色地砖铺设，同时加入书房和餐厅，打破了传统的空间规划理念，给人耳目一新的感觉。

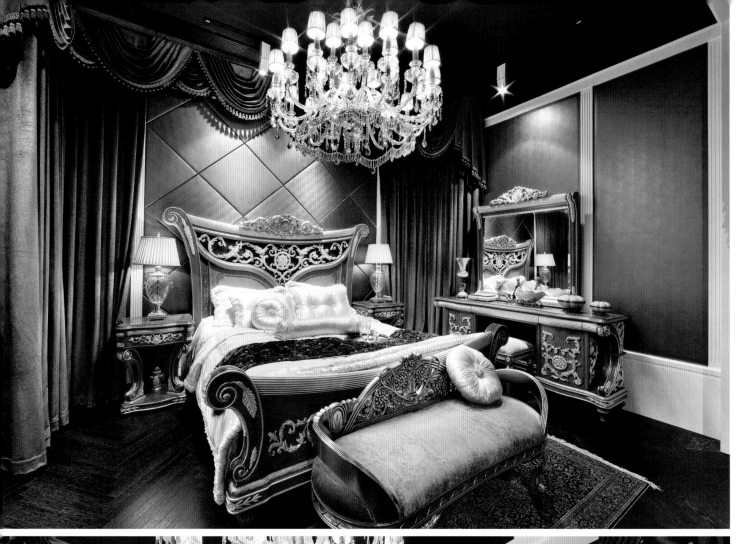

卧室：

  这间卧室不但具备休息功能，同时还加入了餐厅，使整个空间更加饱满。白色的床品和吊灯相呼应，黄棕色的床头和餐桌、酒柜在色彩上保持一致，茶黑色的窗帘象征性地起到规划空间的作用，拉开窗帘使空间更加连贯，拉上窗帘又创造了一个私密安静的休息空间。

卧室：

　　黑色吊顶和地砖给空间营造出沉稳踏实之感，蓝色窗帘帷幔和柔软舒适的床品增添了温馨的气息。卧房空间走向较长，于是在床的对面增加了书桌和书柜，两者居于一室，方便主人阅读、办公和休息。

PROJECT NAME | 项目名称
卡纳湖谷李公馆

LOCATION | 项目地点
重庆

AREA | 项目面积
340㎡

DESIGN COMPANY | 设计公司
品辰装饰工程设计有限公司

DESIGNER | 设计师
庞一飞、李健、殷正毅

MAIN MATERIALS | 主要材料
实木木地板、石材、木作护墙板、墙纸、金箔、手工砖等

SOFT DECORATION DESIGN | 软装设计
程静

A PSALM
如诗的礼赞

DESIGN CONCEPT | 设计理念

Combined the luxury of French classical palace with modern fashion elements, this project presents gorgeous and noble sense of Versailles palace. The delicate and rich line board carvings of the façade wall whose bulges are gilded exquisitely, present a stereoscopic and subtle artistic vision. Wine red velvet furniture and carpet are collocated in the plane space to manifest the luxurious manner. The tall space in the first floor is magnificent, which forms the welcoming manner. The spacious and square living room presents graceful luxury and nobility. The elegant curves on the plain white spiral stair soften the flavor of the space, where you can step up following the huge cascading chandelier. The quartz brick floors with shiny side in public areas have glittering texture like crystal, collocated with the selected carpet and refined furniture, presenting a magnificent and luxurious manner of the mansion instantly. Cohering the noble temperament and exuding the luxurious flavor, this is the French palace style.

本案设计糅合了法式古典宫廷的奢华与当代的时尚元素，呈现出凡尔赛宫殿式的金碧辉煌与尊贵感。立面墙体细腻而丰富的线板雕花，凸起处被细腻地以金箔贴绘，呈现立体而细微的艺术视觉感。平面空间搭配酒红色的丝绒家具与地毯，不错过任何彰显奢华气度的表白。一楼高挑空间，大气恢弘构成迎宾气度，宽敞方正的客厅呈现雍容的奢华贵气。雅白色旋转楼梯，优雅的曲线糅合了空间韵味，可依循直泻而下的大型水晶灯拾阶而上。而公共领域的亮面釉花纹石英砖地面，有着仿若水晶般晶莹透亮的质感，搭配精心挑选的地毯与精制家具，大宅的磅礴富贵气势立刻展现。凝聚贵族气质，散发豪奢风情——法式宫廷风格如是。

SPACE PLANNING　|　空间规划

为了在已有的空间框架中创造出与众不同的设计感，设计师通过垂直面的层次变化，增添视觉上的多样性和趣味性。通过丰富的材料、简约的空间线条，对整个空间进行整合和切割，营造了琳琅满目的效果。设计师以功能规划空间：一楼是主要活动区域，有客厅、会客厅、餐厅等空间，还包括一个下沉庭院；二楼是休息区，有主卧和客卧等；放松区则有休闲室和雪茄吧。

客厅：

挑高客厅、宽阔空间的设置让设计师发挥得更得心应手。整个空间设计极具张力，设计师集合窗帘原本的意形，垂泻而下的线形，在营造大气磅礴氛围的同时，也赋予空间以美的感官享受，让身处这个空间的人不自觉发生情绪上的感染，震慑于空间呈现的气势。

餐厅：

　　餐厅中垂泻而下的水晶吊灯造型优美，好似含苞的花朵，流动光影间传递空间的气质，展示出瑰丽奢华的贵族风范。环绕方形餐桌的丝绒座椅以舒展静默的姿态，共同缔造了空间强烈的秩序感和韵律感。略带冷淡的青色在光耀华贵的空间中更显沉静优雅，带出油画般的视觉体验。

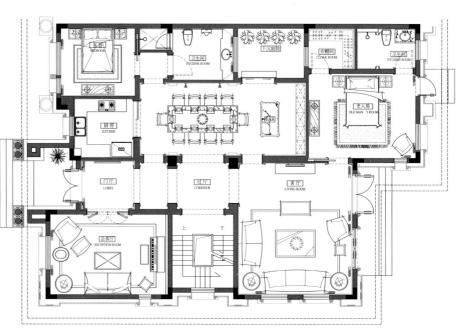

书房：

　　书房区域的吊顶以带有装饰艺术风格的线条勾勒出八边形造型，规律排列，间隔有致。整体设计方正大气，挂饰与摆件相映生辉，墙壁线条曲直相生，图案与光影交汇，展示了空间动态的美感。

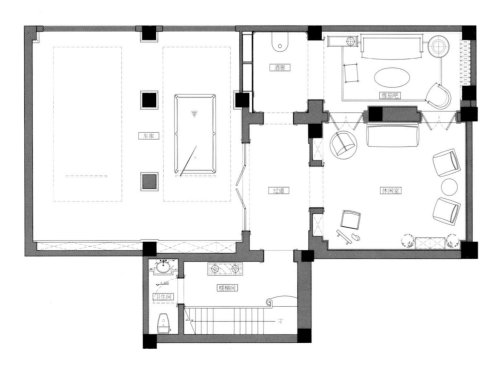

主卧：

　　主卧空间以白色为基底展现大方明亮，配以金色道尽华丽尊贵，配以酒红诉说古典高雅，加上细腻丰富的线板雕花形式将奢华在此定格。从装饰细节的质感着手，循序渐进转化为构成空间的符号和线条，连贯成立面色彩与姿态，缔造整体空间的流畅，空间呈现极致内敛的奢华感。

PROJECT NAME | 项目名称
中星红庐70#

LOCATION | 项目地点
上海

AREA | 项目面积
942.8m²

DESIGN COMPANY | 设计公司
上海鼎族室内装饰设计

DESIGNER | 设计师
吴军宏

PHOTOGRAPHER | 摄影师
三像摄 张静

NOBLE STYLE IN BRITISH MANSION

贵族格调 英式豪宅

DESIGN CONCEPT | 设计理念

This is a British style mansion. The designer uses a large amount of dark clapboard, pillars in different modeling and lines and small ornaments in various patterns to create a pure British mansion, striving to bring people back to the 19th Europe to fully feel the calmness, composure, grace and luxury of the British nobilities. The sedate tone presents a fresh smell, which makes the space rich. Combining with the retro temperament makes the space more magnificent. The designer expects this house can bring visual shocks. So details such as aesthetic curves, fluent lines and complicated carvings, add many highlights to this residence. In addition to pay attention to the decorative effects, the designer restores the distinct temperament of British nobilities by modern methods and materials, which endows the British mansion with classical and modern effects.

本案的设计风格定位为英伦风格，设计师采用了大量深色的护墙板、不同柱式的造型和各种花式的线条及小装饰，打造了一套纯正的英式风格的豪宅，力图把人们带回到19世纪的欧洲，去充分感受英国贵族的从容、淡定和雍容华贵。沉稳的格调中透露着鲜活的气息，使空间更加丰满，与复古的气息相结合，凸显出整个空间的大气。设计师希望这套房子能带来视觉上的震撼，唯美的曲线、流畅的线条、复杂的雕刻等细节的处理为整个住宅增添了不少亮点。在注重装饰效果的同时，设计师用现代的手法和材质还原了英式贵族特有的气质，使这个英式豪宅具备古典与现代的双重审美效果。

SPACE PLANNING | 空间规划

　　一层主要以客厅、餐厅为主要功能，同时设置了一间套房作为父母房，方便使用。设计师在设计上力求空间的宽敞和通透，使之更符合现代人的观点。早餐厅、中餐厅和西餐厅围绕着西式厨房，让三者的空间互动性更强。在客厅上挖出一块空间作为起居室，不仅让功能得到满足，也使整体空间更为宽敞、气派。二层对主卧的更衣室作了调整，使主卧空间更为完整、开阔。因为英式坡顶斜度很大，客厅上方挑空空间无法全部封合，故封合部分作敞开式书房和二楼家庭起居室。地下室以休闲娱乐功能为主，调整了视听室的位置，充分利用下沉庭院的采光和空气流通，让地下室空间更为宽敞、明亮，设置了桌球区、视听区、酒吧区、酒窖区、雪茄区、棋牌室和水疗健身等功能区。

客厅：

　　精致的大型水晶吊灯，搭配华丽的枝型造型，营造出典雅的氛围。门窗上半部分多为圆弧形，并带有花纹勾边。壁炉上方陈设着各种艺术品，烘托出客厅的豪华效果。完美的点、线和精益求精的细节处理，使英式家居给人一种惬意、舒适的触感。

玄关：

　　一楼玄关的位置十分重要，有着承上启下的作用。整体空间的色调、楼梯扶手、护栏、玄关桌上的饰品等都强调了英式古典元素的唯美，精美的装饰画打破了木质墙面的沉闷。空间墙面的线条边框走势英朗、精致，在入口处就给人营造一种大气磅礴的氛围。

餐厅：

以木格天花的中心为中轴线，延伸至吊灯、水晶烛台、餐桌长巾，结构分割清晰明了。两边各有一个餐边柜，镜像的摆设对称铺陈，而墙面上的雕刻装饰镜与装饰画又有所不同，细节丝丝入扣，将英国人特有的严谨发挥到极致。

书房：

　　人字拼木地板、造型窗帘盒、平行线条的吊顶等功能不同的材质搭配，使这个空间不大的书房看起来有条不紊。英式高柜，无论作为书柜还是展示柜都相当实用优雅。沙发椅与书桌搭配，让书房流露出贵族气息，整张牛皮地毯增添了几分刚毅。

主卧：

华丽的吊灯，挑高的内凹格子天花，线条感十足，给人一种视觉上的美感。经典的四柱床顶部曲线造型增加了柔美的质感，床的顶部足够挑高，放置四柱床，效果最佳。羊毛地毯给屋主带来温暖、舒适的感觉。

150　BRITISH & FRENCH STYLE 英法风格

PROJECT NAME | 项目名称
沈阳"仙林金谷"别墅样板房

LOCATION | 项目地点
辽宁沈阳

AREA | 项目面积
400㎡

DESIGN COMPANY | 设计公司
上海无相室内设计工程有限公司

PHOTOGRAPHER | 摄影师
张静

MAIN MATERIALS | 主要材料
雕花护墙板、复古壁纸等

DESIGNER | 设计师
王兵

REAPPEAR THE NOSTALGIC STYLE AND ELEGANCE OF DOWNTOWN ABBEY

再现"唐顿庄园"的复古与优雅

DESIGN CONCEPT | 设计理念

Compared with other European classical styles, connotation and low profile is one of the most obvious features of British classical style. Whether the British gentleman who brings a long black umbrella with a tobacco pipe, or a graceful and retrained British veteran movie star, exudes a noble temperament casually which is not showy nor glaring, pervading slowly like the mellow wine. The designer held this feature to settle the dignified and nostalgic tone in the space. Brown carved panels and retro wallpapers were adopted on all the walls, combined with elegant furniture and delicate chandeliers, making each room become beautiful and dreamy like the movie scenes.

　　与其他欧美古典风格相比，含蓄与低调可以说是英式古典风格最为明显的特征之一。无论是拿着黑色长雨伞、叼着烟斗的英伦绅士，还是婉约优雅的英国老牌电影明星，他们总是在不经意间散发着贵族气息，这气质又绝非是张扬或是耀眼的，只在举手投足间如陈酒缓缓弥散开来。设计师首先抓住这一特点，将空间定下凝重怀旧的基调，所有的墙面都采用褐色雕花护墙板与复古的壁纸装饰，结合优雅的家具、精美的水晶灯，使每个房间几乎都如电影场景般唯美梦幻。

SPACE PLANNING ｜ 空间规划

开放式的客厅排列多扇对称并列的拱窗，因此墙面即使全部用护墙板包覆也并未使空间显得暗淡，反而由于引入明亮的自然采光，带来一种和谐的秩序性的古典美感。挑空的餐厅里，墙面全部用护墙板结合壁纸装饰，落叶黄夹杂铅灰色花纹的壁纸恰到好处地带来怀旧气息。同时在卧室和地下室的休闲空间里，墙面继续延用类似的方式来处理，营造出特有的低调、含蓄、典雅的氛围。

客厅：

墨绿色格纹扶手椅、黑色皮沙发、方几，都具有安妮皇后时期亲切、实用、大方的特点，华丽的海蓝色丝绒长沙发则带有维多利亚时期的印记。而一侧的中式瓷瓶台灯，茶几上精美的雕花酒具，复古的铜扣收纳箱和银制烛台，以及壁炉上方的座钟、印花大瓷盘，于怀旧中渗透着几分贵族气息。

餐厅：

　　餐厅墙上悬挂的印有贵族家徽图案的丝毯，精致的餐具与花艺则让人宛如置身英国古堡中。随处可见的古典或现代题材的油画，一盏盏造型各异，经过做旧处理的铜架水晶灯，都将英式古典风格这份独特的雍容气质推向高潮。

地下室：

在《唐顿庄园》里，精美绝伦的吊灯，闪闪发光的水晶酒杯、银制的餐具、精美的瓷器，给人留下了深刻的印象。对于善于享受生活的人们来说，这些恰恰体现出生活的品位和审美。设计师在陈设上既保留生活气息又不失装饰效果，每件都由设计师精心挑选搭配，营造出一幕幕华美、高雅的生活场景。

主卧：

　　雕刻细腻、框架镀金的湖绿丝绒软包的床架和扶手椅，紫罗兰色的丝绒床尾凳，色彩绚丽丰美，同样也是维多利亚风格的展现。这些不同时期的风格经过巧妙搭配，完美呈现出英式风格的精彩与魅力。

PROJECT NAME | 项目名称 福建龙旺康桥丹堤 D4户型	**DESIGN COMPANY** | 设计公司 陈铌设计	**DESIGNER** | 设计师 陈铌
LOCATION | 项目地点 福建福州	**AREA** | 项目面积 502m²	**PHOTOGRAPHER** | 摄影师 施凯
MAIN MATERIALS | 主要材料 雅士白大理石、灰木纹大理石、白色木饰面、拼花马赛克、拉丝不锈钢、银镜、硬包、墙布等		

THE RHYME OF WHITE POND LILY

白莲花的诗韵

DESIGN CONCEPT | 设计理念

Considering the ideal scenes, the designs of this project continue and pursue the poetic conception and state of the architecture, striving to give deep impressions on the temperament. The interior atmosphere is graceful and luxurious, which brings a magnificent manner on visual sense and a luxurious and comfortable living experience. French columns, carvings and lines are adopted on the details with clear and refined crafts, presenting a romantic and elegant style. Based on profound and white panels, delicate and gorgeous neo-classical furniture is used in the house. Magnificent sofas and drawers are decorated with classical details and embellished with exquisite blue and white porcelain, which presents a French royal taste with more artistic connotations.

出于理想情景的考虑，本案在设计上延续与追求建筑的诗意、诗境，力图在气质上给人深度的感染。室内氛围偏于优雅与华贵，视觉上带来恢宏的气势，豪华舒适的居住体验。细节处理上运用了法式廊柱、雕花、线条，制作工艺精细考究，呈现出浪漫典雅风格。以浑厚而又洁白的护墙板为基底，搭配精致而华丽的新古典家具。例如气势恢宏的沙发组合、大肚斗柜，以抢眼的古典细节镶饰，加上精美的青花瓷点缀其中，多了些艺术底蕴，呈现法式皇族般的品味。

SPACE PLANNING | 空间规划

偌大的别墅合理有序地规划了五层空间，最大化地满足屋主对生活的所有需求。地下室是集娱乐、休闲于一体的功能区域，富有格调的品酒区、雪茄吧，私家定制的恒温酒窖珍藏着陈年佳酿，健身房、棋牌室、桌球室、影音室等，足不出户也能满足屋主所有兴趣爱好的需求。拾级而上，一层以公共礼仪空间为主，正式的客厅，浪漫精致的餐厅，开放式的厨房外加一间双亲房。白色的法式楼梯在墙面使用镶镜处理，让空间变得更开阔。二层和三层以休息区域为主，二层分布着钢琴房、次卧和儿童房，三层则是主卧套房，宽大的露台设计能更充分地领略窗外四季美景。顶层面积相对较小，于是设计成独立的书房和办公区间，安静的空间适合在此办公和阅读。

客厅：

米黄色体量大的沙发和茶几尽显法式优雅的浪漫，沙发线条柔美，缎面软包舒适有度，印花靠枕则又略带法式小清新。蜡烛式的水晶透明吊灯与墙面壁灯相呼应，天花上环绕着精美的雕花，细腻富有设计感，茶几的蓝色摆件又起到空间色彩的点缀作用。

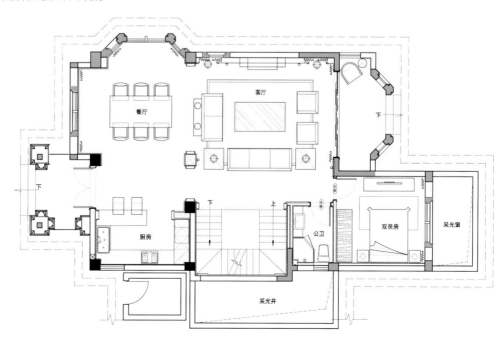

钢琴房：

  一台白色的钢琴精置于窗边，白色的绒毛地毯显示出女性的优雅柔美特质，休闲区的贵妃躺椅和高背单椅、脚榻，采用珍贵的缎面，尽享法式舒适、典雅的风情。墙面孔雀元素的装饰画，高贵富丽，搭配蓝白花色壁纸，协调又统一。

品酒区：

错落有致的方格恒温酒窖珍藏着屋主喜爱的名酒，金色和黑色元素的大面壁画与窗帘、吊灯在色彩上相呼应，高脚的金色吧台椅线条感强，给人干净利落的质感，搭配白色的吧台与空间墙面色彩相协调，同时又颇具浪漫的休闲情调。

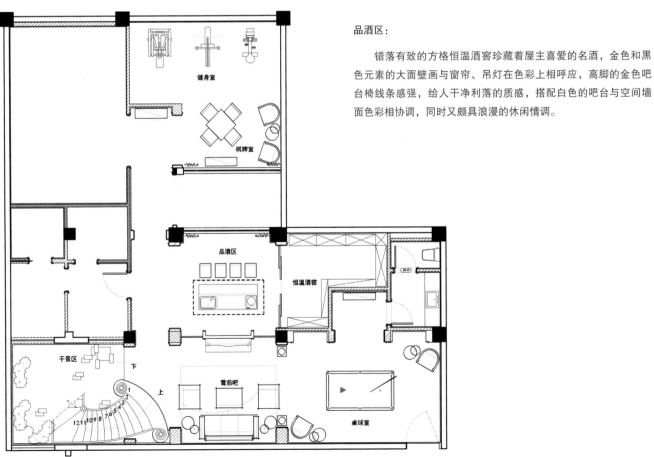

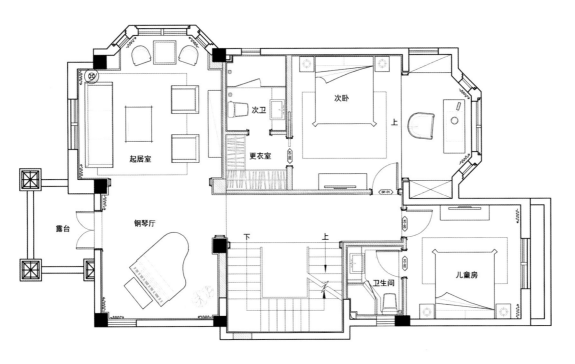

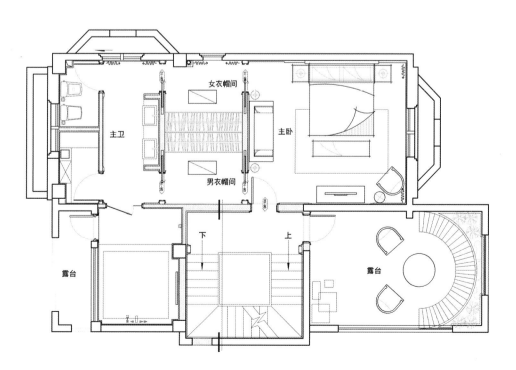

主卧房：

　　白色床品与蓝色床头的碰撞，彰显出清新、浪漫的气息，墙面使用蓝白搭配的壁纸，同时与窗帘互为一体。床头柜、电视柜和窗边的梳妆桌，都有着精美的雕饰，电视柜上的青花瓷摆件，又散发出中式的韵味，同时在色彩上与空间保持一致。

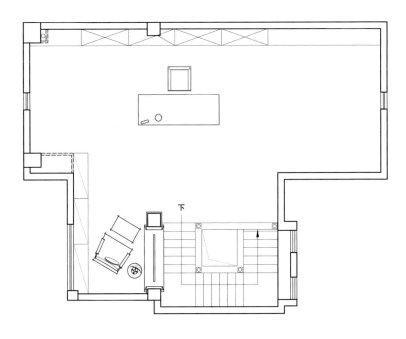

190　BRITISH & FRENCH STYLE　英法风格

PROJECT NAME | 项目名称
苏州水岸西式秀墅

LOCATION | 项目地点
江苏苏州

AREA | 项目面积
344m²

MAIN MATERIALS | 主要材料
蒙马特灰大理石、原色油面崖豆木地板、梧桐喷砂实木拼、白色钢琴烤漆、明镜、灰镜、蓝色油性平光漆等

DESIGN COMPANY | 设计公司
玄武设计

SOFT DECORATION LAYOUT | 软装布置
杨惠涵、张禾蒂、沈颖

SCRIPT | 文字
程歆淳

COOPERATING DESIGNERS | 参与设计
黄书恒、苏幼君

PHOTOGRAPHER | 摄影师
王基守

DEEP CHANT OF BLUE

海蓝深咏

DESIGN CONCEPT | 设计理念

"When time stands still, the scenery is condensed in the traveler's view, while the only sea blue spreads from the center, paving the background of the whole city." If the world moves its shapes and shadows, for example, removing Venice, the water city, to China, then how will the landscape be? The two cities which occupy respectively the east and west in the world, have the same bizarre imagination on "water". Sherwood Design Group uses the images of fresh blue sky and navy blue ocean in Venice, collocated by gentle and soft Victorian style, which makes the residents here experience the exclusive and deep amorous feelings of this project through the changes in colors and lines.

"当时间静止，风景凝结于旅人的视野，唯一抹海蓝自中心漫开，铺就整座城市的底蕴。"假如世界移形换影，将水都威尼斯迁移至中国，会是如何风景？分据世界东西的两座城市，同样对"水"有着奇异的想象。玄武设计用威尼斯的碧蓝天色与湛蓝海洋意象，搭以柔和婉约的维多利亚风，使居住者于深浅变换、线条起伏之间，体验专属本案的深邃风情。

客厅：

　　白色天花和墙体柜，线条感分明，给人干净明快之感，蓝色大面积的印花地毯，与单座沙发椅互为呼应。收藏柜、天花和茶几采用灰面镜装饰，反光材质让色彩在空间得到碰撞与交汇。

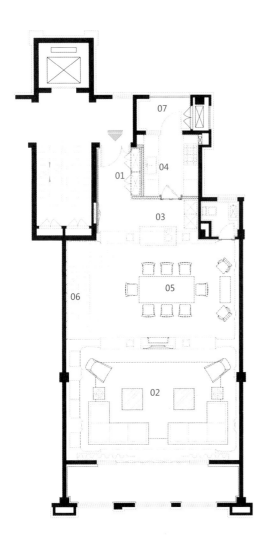

SPACE PLANNING | 空间规划

踏入玄关，大幅沉稳壁色沉淀着访客心绪，彷佛进入高潮之前的低沉乐音，诱人缓步轻移，步入大厅，可见西厨吧台区域的纯白色泽与铁灰镜面，体现强烈的戏剧张力；为扩展景深并消解低梁带来的压迫感，设计者特别利用三座连拱的流畅弧线，借由动态的视觉起伏，纾解空间压力；延续设计主轴，设计者选用深色木皮为楼梯处主色，然于接缝处，嵌入一盏盏小型led灯，光线影影绰绰、随着步履忽隐忽现，仿若大隐于苍穹的点点星光，为阆深的壁面点染几分趣味，也削减了色彩过于沉重的感觉。

上至二楼，大面锻铁金漆扶手蕴藏浓厚西式韵味，与实木地板、水晶灯等，维系着空间的大气磅礴。行至主卧，设计者在一片隽蓝之中，运用中国青花瓷、湖水绿等色泽，创造多层次视觉变化；次卧铺就小碎花壁纸，床背板以铆钉排出流利图腾，一如卫浴间的黑白几何分割，均在古典工艺与现代美学之间，凝炼收放自如的平衡美感。

餐厅：

餐厅利用餐椅与灯具，体现素白、浅灰与碧蓝的色彩游戏，侧边一只复古壁炉，呈现精致的英伦风韵，与客厅电视墙合而为一的精巧设计，亦可见营造焦点、避免对象散乱的匠心。

01 PORCH	01 玄关
02 LIVING ROOM	02 客厅
03 WESTERN KITCHEN	03 西厨
04 CHINESE KITCHEN	04 中厨
05 DINNING ROOM	05 餐厅
06 STAIR HALL	06 楼梯间
07 WORK BALCONY	07 工作阳台

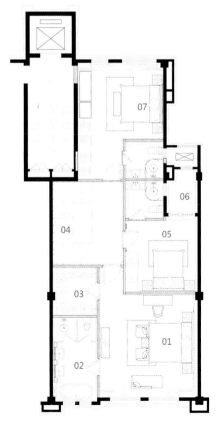

01 MASTER BEDROOM	
02 MASTER BATHROOM	
03 DRESSING ROOM	
04 STAIR HALL	
05 CHILDREN'S ROOM	
06 BACK BALCONY	
07 SECOND BEDROOM	

01 主卧室
02 主卫浴
03 更衣间
04 楼梯间
05 小孩房
06 后阳台
07 次卧室

主卧：

深蓝色窗帘内搭白色轻纱，朦胧轻慢，贵妃椅横卧窗边。床品以珍贵缎面织造，床头墙面白色软包以几何造型规划空间，与天花形状相呼应。浅青色的沙发椅和黄色插花，营造了清新浪漫的气氛。

PROJECT NAME | 项目名称
宁波世茂海滨花园

LOCATION | 项目地点
浙江宁波

AREA | 项目面积
850m²

DESIGN COMPANY | 设计公司
上采国际设计集团

DESIGNER | 设计师
蔡文亮

DEDUCING ARISTOCRATIC PALACE TEMPERAMENT

演绎宫廷式贵族气质

DESIGN CONCEPT | 设计理念

This is a gorgeous household space. Through the configuration of furniture, the collocation of colors and lines in the space, the designer endows the majestic space with aristocratic palace temperament. The magnificent Roman columns gorgeous crystal droplights, rich fabrics and various soft decorations deduce the luxury and elegance of the home into a flowing scenery line, which makes every visual view full of the enjoyment of high quality. The designer uses modern processing technique to present the classical palace temperament, abandons complicated furnishings, uses fashionable, elegant and gorgeous design languages to create a romantic home full of noble spirits.

这是一个富丽堂皇的家居空间，设计师通过家具的配置、色彩的搭配及空间线条的勾勒，赋予这个大气磅礴的空间以宫廷式贵族气质。气势十足的罗马柱、华丽的水晶吊灯、丰富的布艺及各种软装饰等家的华贵与典雅风范演绎成一通流动的风景线，让每一个视觉角度都充满着高品质的享受。设计师透过现代的处理手法，将宫廷式古典气质彰显得美轮美奂，摈弃了繁复的修饰，运用时尚、高雅、华丽的设计语汇将空间打造成一个充满贵族神采的浪漫之家。

SPACE PLANNING | 空间规划

本案中设计师运用丰富、流畅的设计词汇构建了一幅家居梦想的蓝图，融入一系列古典的元素，通过具有古典韵味的华贵布艺、家具，大放光彩的水晶灯饰，各式工艺品摆件、装饰画，色彩鲜艳的花卉植物等来装点空间。同时，在空间构造上，设计师擅用罗马柱、拱形门框、古典线板及各式天花板为空间规划出鲜明的层次美感。设计师透过各种元素之间的糅合、交叠，创造模拟化的场域、情境，将宫廷式贵族生活的画面搬进现代人居住的空间，追寻历史的余味，享受"坐拥天下"的富贵之感。当然，这一切都是为了营造家之本身的温馨与浪漫。

客厅：

繁华的布艺软装饰是本空间一道靓丽的风景，以其温润、华美、端庄、古典吸引人的视线，贯穿于空间区域的每一个设计表达之中。地毯质地上乘，以墨绿色、咖啡色的大幅花卉图案作为装饰，既与空间的大气磅礴相呼应，又呈现出自身的端庄与清雅。而沙发上咖啡色、墨绿色抱枕则与地毯完美搭配，自成一体，其简约的设计外观给人一丝清新、明快之感。

餐厅：

与客厅不同，餐厅以其细腻、秀丽为突出，与餐厅整体淡雅的环境融为一体，尽显"小家碧玉"式的秀雅与美丽。上下方的两张红色的椅子格外引人注目，与花卉图案的地毯、孔雀图案的椅子彼此映衬，相得益彰。白色水晶吊灯散发着柔和的光线，浪漫、温馨的气息氤氲四周。

会客厅：

　　整个空间色彩搭配深浅有度，给人不一样的视觉感受。作为家中的会客室，设计师选用了偏淡雅的色彩搭配，其中米白色和咖啡色作为空间的主打色，配合沙发的绿色、桌几的蓝色，更丰富视觉的层次感，同时水晶灯温馨的光线下使空间洋溢着舒适而芬芳，沉静而智慧的气息。

主卧：

放眼望去，本案设计师将布艺的繁华之貌演绎到极致，无论是窗帘、地毯、抱枕还是床品，都以其不同的色彩、纹样、材质，彰显出丰富的内涵，每一处都散发出浪漫、温馨的气息，将古典与时尚巧妙结合，诠释着不一样的唯美视觉。

BRITISH & FRENCH STYLE 英法风格

PROJECT NAME | 项目名称
大连龙润 橡树庄园二期别墅

LOCATION | 项目地点
辽宁大连

AREA | 项目面积
500m²

DESIGN COMPANY | 设计公司
深圳市砚社室内装饰设计有限公司

DESIGNER | 设计师
姚海滨

MAIN MATERIALS | 主要材料
石材、马赛克、墙纸、地毯等

CULTURED AND REFINED HOME

博雅之轩

DESIGN CONCEPT | 设计理念

Surrounded by armored concretes, contemporary people naturally need a space to ease their minds and enjoy their lives, a place where they can stay alone when upset and can have fun with family when clam, and a place where they can escape from the bustling world. Step into the foyer, one will feel open and clear suddenly. The pure white arched doorway is extended to the entrance hall, which connects the living room, reception room and staircase, playing a central role. The foyer and entrance hall adopt marble walls and floors, which is concise and generous. This kind of open structure designs make the space more transparent and coherent. Walk slowly and move forward, the living room has no solemn and rigorous feelings as the foyer, but only keeps a classical sense subtly, such as the black closet with gilt-edge in Chinese style. The villa mainly uses light colors, such as light blue, lake blue and off-white, reflecting a graceful atmosphere. The reception room continues the style of the living room, with light blue and white fabric sofas and tea table, making people feel comfortable and relaxed. The gathering dining space is particularly in need of a calm and pure ambience so that the designer uses pure blue wallpaper as the setting with flower murals on it. The red color of the dining table is as delicate as the color of the walls of Qin and Han Palaces. The brand porcelains on the table are overflowed with elegant taste of celebrities.

当代人在钢筋水泥的包围下，自然更需要一处可以沉淀心灵、纵情视听的空间，能够于烦躁时个人独处、平和时家人同欢的场所，使人遁逃于纷乱世事之外。步入玄关，顿时觉得豁然开朗，纯白的拱形门洞延伸到门厅，贯连着客厅、会客厅以及楼梯，扮演中枢的角色。玄关与门厅采用大理石材质的墙面和地板简洁而大气，这样的开放结构设计使得空间更为通透连贯。缓步轻移，可见客厅空间一反玄关的庄重严谨，只在细微处维持古典感——如黑漆金边的中式立柜；在色彩上以各种清淡色为主，如浅蓝、湖蓝、米白等，彰显雅致的空间氛围；会客厅延续客厅的风格布置，淡蓝、白的布艺沙发、茶几，让人倍感舒适放松。欢聚的用餐空间，特别需要沉稳、纯净氛围，设计师用纯美蓝壁板为底，附上花鸟壁画；如秦汉宫墙红般的餐桌，细腻色泽与上方的品牌瓷器，盈溢着典雅的名流品味。

客厅：

　　层层递进的天花设计，线条感十足的墙面设计，对称的空间格局，使这个白色空间越发显得高贵、大气。白色的沙发与整个空间完美融合，抱枕上跳跃的蓝色和茶几上鲜艳的粉色增添了几许活泼灵动。立柜、茶几、水晶灯、摆件等，于细节处取胜，展示着精致的雕刻技术。

会客厅：

　　秉承客厅的装饰风格，浅蓝色的沙发、单椅，搭配蓝色纹理的地毯，给人一种轻松自在的感觉。大面积的开窗设置，天蓝色的薄纱，微风袭来，营造了一种飘逸的动态美。

餐厅：

　　方形的红木餐桌与天花遥相呼应，挑高顶面，烛台吊灯垂感十足。淡蓝色的墙面上，写意的鸟语花香，不仅能增进食欲，更是视觉盛宴。古铜镜做工精细，透露出空间的古色古香。餐具精致，摆放有序，可见屋主极其讲究生活的品质。

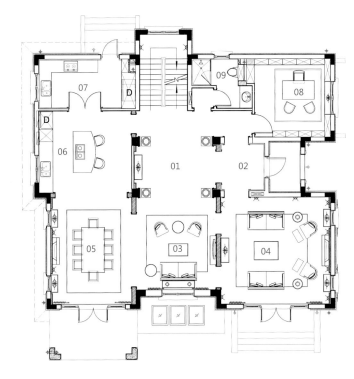

01 FOYER
02 PORCH
03 FAMILY ROOM
04 LIVING ROOM
05 DINNING ROOM
06 WESTERN KITCHEN
07 CHINESE KITCHEN
08 STUDY
09 PUBLIC BATHROOM

01 过厅
02 玄关
03 家庭厅
04 客厅
05 餐厅
06 西厨
07 中厨
08 书房
09 公卫

玄关：

　　连续的拱形门，延伸了空间深度；挑高的白色吊顶，拉伸了空间的高度；精致的水晶吊灯散发出温暖的光照耀在反光的大理石地板上，折射出一个时尚、有质感的空间。天花、门形、地板大量使用的几何图案，创造出视觉美感。

SPACE PLANNING | 空间规划

根据屋主的生活及工作需求，这个别墅被分为三层，地下一层，地上两层。地下一层主要有女主人工作区、休闲娱乐区及设备区，还有一个车库。一层为家庭生活的公共区域，二层主要为家庭成员的私人生活休息区，包括主卧、男孩房、女孩房等。这种简单合理的空间规划，为屋主及家人提供了极大地便利。

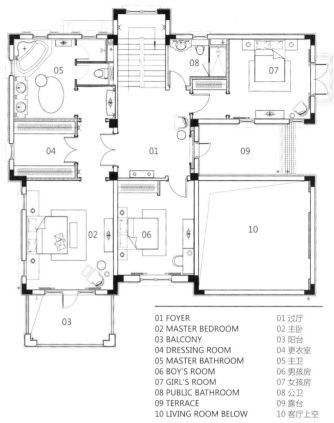

01 FOYER	01 过厅
02 MASTER BEDROOM	02 主卧
03 BALCONY	03 阳台
04 DRESSING ROOM	04 更衣室
05 MASTER BATHROOM	05 主卫
06 BOY'S ROOM	06 男孩房
07 GIRL'S ROOM	07 女孩房
08 PUBLIC BATHROOM	08 公卫
09 TERRACE	09 露台
10 LIVING ROOM BELOW	10 客厅上空

01 FOYER	06 AUDIOVISUAL ROOM	01 过厅	06 视听室
02 FAMILY ROOM	07 LAUNDRY	02 家庭厅	07 洗衣间
03 HOSTESS WORKING AREA	08 EQUIPMENT ROOM	03 女主人工作区	08 设备间
04 WATER BAR	09 MAID'S ROOM	04 水吧	09 工人房
05 CELLAR	10 GARAGE	05 酒窖	10 车库

PROJECT NAME | 项目名称
尚湖·江南府邸别墅样板房

LOCATION | 项目地点
江苏常熟

AREA | 项目面积
508㎡

DESIGN COMPANY | 设计公司
上海全筑建筑装饰集团股份有限公司

ENTIRE SOFT DECORATION | 整体软装
苏州美哲艺术软装有限公司

MAIN MATERIALS | 主要材料
象牙白浑水漆、樱桃木饰面、米色墙纸、香槟金箔、黑金花石材、实木地板、仿古砖等

PHOTOGRAPHER | 摄影师
金啸文空间摄影

A CLOUDY MANSION

云·邸

DESIGN CONCEPT | 设计理念

The elegant and romantic atmosphere is pervaded in this show flat with French neo-classical style like a melodious violin song. It is the wonderful atmosphere embellished by the gorgeous chandeliers, delicate tea pots and comfortable sofas, as well as spatial connotations created by the fresh ivory white, elegant dark green and noble gold. The gorgeous fabric and delicate carvings on the furniture embody the sense of luxury. The decorative ornaments of copper with porcelain and the retro floriculture become the highlights of the space. Classical romantic flavor and modern people's pursuit of delicate life are combined in the overall space, which is luxurious, elegant, fashionable and modern, creating an ideal residence where the elites can nurture their bodies and souls.

　　如沐一曲悠扬的小提琴乐，典雅、浪漫的气息在这套法式新古典风格的样板房中缓缓弥漫开来，那是华美的水晶灯、精致的茶盏、舒适的沙发烘托的美好氛围，更是清丽的象牙白、雅致的墨绿、高贵的金色所营造的空间意蕴。家私中华丽的布面与精致的雕刻如影相随，尽显雍容华贵之势。铜配瓷的装饰摆件和复古的饰品花艺，成为空间的点睛之笔。整个空间将古典的浪漫情怀与现代人对精致生活的追求相结合，兼容华贵典雅与时尚现代，缔造出一处精英人士修养身心的理想居所。

SPACE PLANNING　｜　空间规划

　　空间布局突出轴线上的对称，以营造恢弘的气势。在原有硬装的基础上，软装可以帮助解决空间规划的问题，使空间的线条显得更加流畅，突出整体感。因此，设计师通过了解屋主对不同空间的功能要求，基于实用性、奢华性以及舒适性，对整个空间进行了合理的规划。

客厅：

独特的水晶吊灯与挑高的现代椭圆金色天花交相辉映，使整个客厅看起来非常宽敞。线条流畅的法式新古典风格的沙发、单椅、茶几等，以工整的布局安放，随处可见的绿植、摆件，彰显了屋主优雅的生活情趣。门以及橱柜上的半弧形几何图形和石膏线的雕刻尽显屋主对高品质生活的追求。

餐厅：

　　挑高的圆形吊顶与红木圆形餐桌相互呼应，吻合新古典对应的特质。半开放式的设计，使空间更开阔。卡其色的花纹餐椅，为空间增添了几许华贵气息。流畅的线条与几何图形的搭配，增加了空间的质感。

台球室：

  挑高的天花吊顶、铁艺吊灯、台球桌，均选用方形造型，在视觉上达成一致性，线条感十足，轮廓清晰，塑造出一个时尚、有质感的休闲运动空间。几何图形的大理石地板，给整个空间注入了活泼青春的气息，与空间的功能性相符合。

书房：

　　书房的布局简单工整，色彩和谐，达到了整齐划一的效果。大幅地图、轮船模型等装饰与屋主热爱旅游、冒险相契合。圆形的印花大地毯，用其沉稳的颜色，以恬静的方式带给屋主心灵的放松。

主卧：

床帘、窗帘、窗纱等的直线设计与墙面的线条设相互呼应，高雅干练。床品的选择以舒适、贵气为主，床头龛、床头凳、单椅等均有法式特有的精致雕刻，衬以大型挑高的天花，营造出浪漫时尚的氛围。此外，还有一个休闲区，为屋主创造了足够的私人空间。遥遥相望的两端大开窗设计，不仅增强了空间的采光度，也促进了空气的流通。

PROJECT NAME	项目名称	**DESIGN COMPANY**	设计公司
常州中海锦龙湾别墅		上海桂睿诗建筑设计咨询有限公司	
LOCATION	项目地点	**DESIGNER**	设计师
江苏常州		桂峥嵘	
MAIN MATERIALS	主要材料		
大理石、雕花木线、烤漆、布艺等			

DIGNIFIED ENJOYMENT OF LIFE

尊贵生活享受

DESIGN CONCEPT | 设计理念

French style emits a retro, romantic and naturalistic tone, whose most prominent feature is the noble atmosphere. Such "nobility" exudes a humanistic and classical beauty. This style reminds people of manor, piano, ball and bubble skirt. The refined palace style is the "The Peach Garden" that people have been pursuing of. Soft and slender decorative panels, gold and silver foil carved lines, delicate and beautiful ground mosaic interpret the romantic and sumptuous French palace style as if a graceful noblewoman. The elegant and luxurious French atmosphere also needs proper decorative colors, such as gold, purple and red which gently jump within the white color, rendering a soft and elegant temperament. Whether it is abstract art flower pattern or floral monogram, each will make the room more sensible and full of romantic and enchanting feminine colors.

　　法式风格弥漫着复古、浪漫、自然主义的情调，最突出的特征是贵族气十足，这种"贵"散发着人文气息和古典美韵。此风格让人联想起庄园、钢琴、舞会、蓬蓬裙，精致化的宫廷风，是人们一直都在追求的"桃花源"。柔美修长的装饰镶板，金银箔的雕花木线，细腻精美的地面拼花等，无不诠释着浪漫华贵的法式宫廷风，犹如一位婀娜多姿的贵妇人。优雅而奢华的法式氛围还需要适当的装饰色彩，如金、紫、红，夹杂在白色的基础中温和地跳动，渲染着一种柔和高雅的气质。无论是抽象艺术花形，还是碎花组合图案，都会使居室变得更为感性，极具浪漫妩媚的女性色彩。

客厅：

　　优雅静谧的蓝色是法式风格的特点之一，浪漫且洋溢着现代的时尚气息，为规整的空间注入活力。融合古典与时尚美感的家具配置、华贵的布艺、精致的水晶吊灯、工艺品摆件、装饰画等，营造出高贵典雅的氛围。空间构造极富线条感，大气且注重细节上的精雕细琢的美。

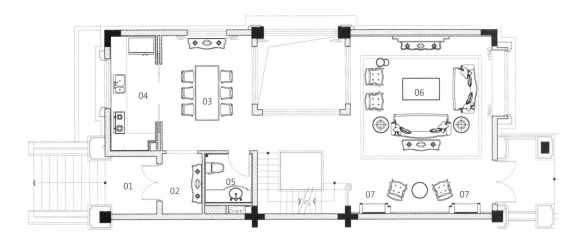

01 VESTIBULE
02 FOYER
03 DINNING ROOM
04 KITCHEN
05 BATHROOM
06 LIVING ROOM
07 SHOW AREA

01 门廊
02 门厅
03 餐厅
04 厨房
05 卫生间
06 客厅
07 展示区

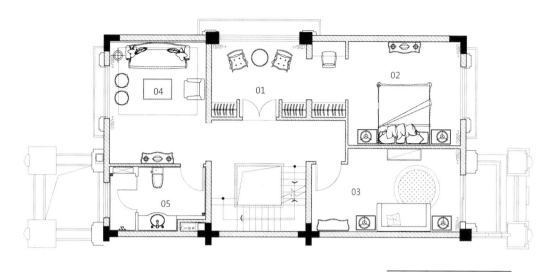

01 FOYER	01 门厅
02 PARENTS ROOM	02 父母房
03 GIRL'S ROOM	03 女孩房
04 FAMILY ROOM	04 家庭室
05 BATHROOM	05 卫生间

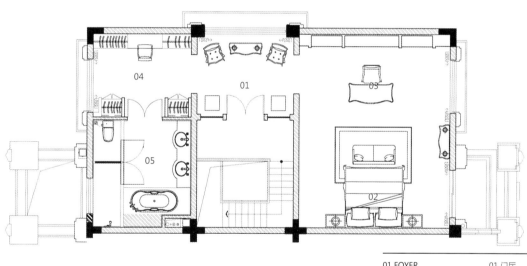

01 FOYER	01 门厅
02 MASTER ROOM	02 主卧室
03 STUDY	03 书房
04 CHANGING ROOM	04 更衣室
05 BATHROOM	05 卫生间

SPACE PLANNING ｜ 空间规划

整个别墅一共有四层，一层为家人活动的公共区，客厅、餐厅、厨房构成主动线。二层有父母房、女孩房、家庭室，较低楼层的安排设置比较合理，方便老人与小孩的生活。三层为屋主的私人空间，包括主卧、主卫、书房、更衣室等，为屋主提供了宽敞且足够的个人生活和工作区域。地下一层为休闲娱乐区，有陶艺区、儿童活动室等，此外还有一个大车库。

餐厅：

餐桌餐椅造型简单复古，纯红木手工制造，展示出低调、内涵的华丽。天花的形状与餐桌相呼应，黑白相间的菱形大理石地板，圆锥形的烛台吊灯，精致时尚，大量几何图形的合理使用营造出一种视觉美感。插花、摆件则让整个空间弥漫着浪漫温馨的气息。

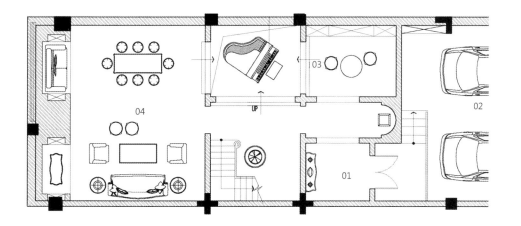

01 FOYER
02 GARAGE
03 CERAMIC ART AREA
04 CHILDREN'S ROOM

01 门厅
02 车库
03 陶艺区
04 儿童活动室

主卧：

  挑高的天花设计层层递进，营造出视觉的延伸感。墙面上宽度不一的长方形框和竖条纹，使空间里有了线条的流动感。床品突出舒适、精致、高贵的特点，各种纹理和图案的搭配，创造出内在和色彩的完美平衡。主卧兼备书房的功能，靠墙的整体书柜与柜子结合的设计，极大地发挥了空间利用率，各种书籍与古玩整齐地摆放在书架上，赏心悦目。

女孩房：

　　丰富的色彩符合儿童天真浪漫的心智和开阔的思想。窗帘和床帘上的红心，墙壁上的热气球挂画，柜台上的小玩具摆件，展示着小女生的梦幻公主梦。

248 BRITISH & FRENCH STYLE 英法风格

PROJECT NAME | 项目名称
布利杰印象巴黎样板房

LOCATION | 项目地点
浙江宁波

AREA | 项目面积
248m²

DESIGN COMPANY | 设计公司
杭州极尚装饰设计工程有限公司

PHOTOGRAPHER | 摄影师
娄俊松

MAIN MATERIALS | 主要材料
木饰面加香槟金箔家具、铜加水晶灯具等

DESIGNER | 设计师
沈肖逸

THE BLUE DREAM IN THE MANSION

庄园蓝梦

DESIGN CONCEPT | 设计理念

Retro and naturalistic flavor is pervaded in the romantic French style, reminding people of the manor, piano, ball and bubble skirt. The refined country style is the "the Peach Garden" which is pursued by people all the time. This project aims at creating a living atmosphere with romantic French flavor which is comfortable, leisurely and slow. Every space owns different roles and expressions. The living room welcomes you into the house like a tall and elegant gentleman, while the dinning room continues the elegance of the living room, bringing a clean and transparent dinning environment. The expression in the master bedroom is steady, calm and elegant, while the daughter's room is a small garden for the lovely princess to enjoy the innocence and happiness.

浪漫法式风格弥漫着复古、自然主义的情调，不禁让人联想起庄园、钢琴、舞会、蓬蓬裙。精致化的乡村风格，是人们一直都在追求的"桃花源"。本案意在营造一个浪漫法式风情的家居氛围，以古寓今，打造不一样的舒适慵懒、随心惬意、慢时光的居家氛围。每个空间都有其不同的角色与表情，客厅犹如高挑儒雅的绅士邀你入室，餐厅延续了客厅的雅致，带来明净透亮的用餐环境。主卧的空间表达是稳重的，沉静明雅的；女儿房则是可爱公主尽享天真快乐的花园小屋。

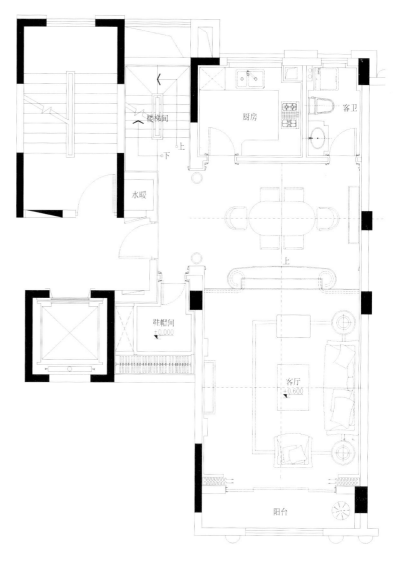

客厅:

  法式家居风格通过温馨简单的色彩及素朴的家具，使整个空间都散发着人文和古典的气息，而舒适、优雅、安逸是它的内在气质。客厅蓝色绒拉扣沙发与直泻而下的蓝绸窗帘让整个空间沉入一片蔚蓝的海洋之中，入室的心情即刻随之安稳沉静下来。地面的蓝色与白色相间的地毯，有着蓝天白云般的干净与自然。室内墙面色调均为米黄色，空间氛围温馨柔和，没有一丝多余的累赘感，更没有一丝不入格调的色彩入室，整体色彩搭配一气呵成，温情浪漫。

SPACE PLANNING ｜ 空间规划

通透而富有层次感是设计师巧妙营造的空间视觉效果，开阔的挑高客厅，沿阶而下的餐厅空间，错落有致，高低相成。同时，选用丰富而大气的软装搭配亮丽的蓝色，在大小不一的格局中各个空间又彼此呼应。浓浓的法式情怀，在一开一合间尽享雅致。

餐厅：

硬装线条细腻优雅，透出法式的精雕细琢。软装陈设萃取法式风格的精髓，家具造型线条柔美，精工细作，富含艺术气息，再配以大量大方不做作的饰品，呈现出普罗旺斯的浪漫，将现代西式生活场景演绎得精致而清新。

卧室：

　　空间色彩继续以米色调搭配蓝色为主，温馨舒适。不论是卧室花瓶里娇艳的花朵，抑或是床头柜上简约的几幅相框摆台，在任何一个角落，都能体会到主人悠然自得的生活和阳光般明媚的心情。

DESIGN COMPANY | 设计公司
北京意地筑作装饰设计有限公司

LOCATION | 项目地点
北京

DESIGNER | 设计师
连志明、张伟、徐辉

AREA | 项目面积
361m²

THE SEA OF PARIS

巴黎的海

DESIGN CONCEPT | 设计理念

As a top-grade property located in the west of Beijing facing the university town of Haidian District and the high-tech talents of Zhongguancun District, the designers target the clients with international education background and overseas life experience as the client community. Under the buildings in the community and landscape environment with French flavor, the designers position this show flat as a French style with relatively strong spatial and color attributes, which fits for the living habits of Chinese people. Designing the foreign-style house into a villa is the spatial design emphasis of this show flat, which owns a distinct theme in Chinese style, a setting that is suitable for the living habits of Chinese people and the effective application of the space, so that this house type becomes one of the types that more clients have selected.

作为北京西部一处面向中关村，高新区海淀大学城的高档楼盘，我们将客户设定为有国际教育背景和有海外生活经验的客户群体。在具有法国风情的小区建筑与景观环境下，设计师将此套样板间定义为具有中国人生活习惯的，空间与色彩属性较强的法式风格。将洋房设计成别墅是此样板间的空间设计重点，鲜明的法式风格主题，中国人习惯的生活方式设置与空间多效性运用使此样板间成为客户选择较多的户型之一。

客厅：

  客厅的色调是柔和淡雅的，灰绿调的墙面与亮白色的墙线穿插排列，奠定了空间的基调，同时在白色的雕花吊顶与白色的地面之间起到承上启下的连接作用，让空间色彩丰富，层次分明，初入客厅，给人安静平和的视觉感。灰调的绿色墙面延至每个空间，偶尔有暗红色的花藤壁纸或软装饰品，远远的看着，犹似巴黎矮墙上开着大朵的蔷薇花，美得不可言说。

餐厅：

　　法式的浪漫风情由客厅延伸至餐厅，色调也彼此呼应。暗红色的花藤壁纸构成了餐厅空间的主墙面，为了减少太过繁复艳丽的压抑感，设计师巧妙运用了镜面元素，精致雕刻的金色镜框亦是亮点，同时透过镜面我们可以看见精美丰富的餐厅场景。

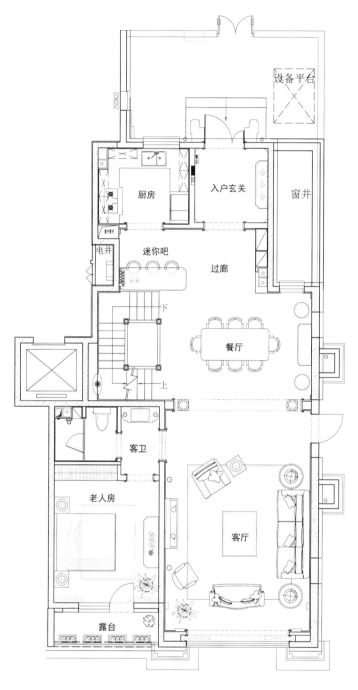

休闲室和书房：

　　延续设计师风格的统一性，书房及休闲室同样采用了灰绿的墙面处理，线条感更加丰富立体，空间氛围更加轻松惬意。一旁静蛰的大钢琴，可以给家人带来听觉的盛宴。同时，与书房相连，知识的增长亦是倍感欣慰。淡淡的墨香中，品位生活的真谛。

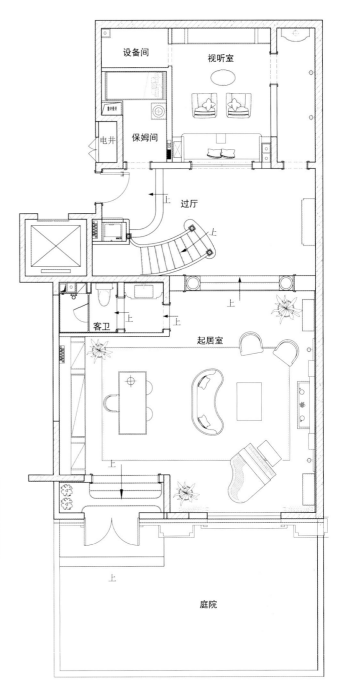

SPACE PLANNING ｜ 空间规划

　　法式风格是大众最为喜欢的一种空间设计语言，浪漫、优雅、高贵且历久呈新。设计师在重新定义空间属性后，开始在空间设计里做足文章，在满足空间领域功能性的同时让室内妆容清新雅致，过目不忘。餐厅与客厅、厨房浑然一体，相互借用，共通互融，开放式的用餐环境让居者没有局促感和压抑感，视线更为开阔明朗。而一楼老人房、书房、更衣室、主卫、主卧动线合理，让此只有300平方米的空间有着极为丰富的层次感。

卧室：

　　舒适柔软是卧室空间给我们传递的视觉信息，柔和的珠光绿软装组合，给人清新自然的亲近感。大气的款式与绸缎的质感，给人更舒心的享受。而磨砂玻璃的运用，扩宽了空间视野，同时在水晶吊灯的闪耀下空间更显朦胧。

BRITISH & FRENCH STYLE 英法风格

PROJECT NAME | 项目名称
龙湖紫宸

LOCATION | 项目地点
重庆

AREA | 项目面积
1700㎡

SOFT DECORATION | 软装设计公司
重庆元禾大千艺术品有限公司

DESIGNER | 设计师
耿波、吴梁祝

MAIN MATERIALS | 主要材料
大理石、真丝、丝绒、仿古镜、古铜等

A MONUMENTAL MANSION

传世大宅

DESIGN CONCEPT | 设计理念

The top-level design team of mansions in the world uses a unique craft of "Chinese spirits with Western techniques" to trigger a series of changes on architectural symbols, which interprets the power, nobility, luxury and mystery of the magnificent flavor in Dang dynasty to the extreme, so does the Dragon Lake which presents it better by its 20 years' experience on building villas. Designers in Esperluette Artwork combine both Chinese and Western cultures to put the romance of French style and the solemnity of the courtyard houses in ancient times on the foundation of the imperial power, creating a spatial atmosphere which is not limited to be a single style. The customized soft decoration designs from a spatial angel of view in large scale interpret the peak of the Dragon Lake single villa, which is an inclusive mansion that inherits the glory of a generation.

世界顶级豪宅设计团队以"中魂西技"的独特打造工艺引发建筑符号的一系列蜕变，把盛世唐风的权力、华贵、隐富、神秘演绎到极致，而龙湖以其20年的别墅打造经验，微至豪厘。元禾大千软装设计师融合中西文化，以九五至尊为根基，将法式的浪漫与古时大院的庄重合为一体，打造出的空间气场将不再仅仅是局限于一个单一风格。以大尺度的空间视角，定制的软装设计，演绎出龙湖独栋登峰造极的高度，一座府邸包容、传承一世荣耀。

客厅：

元禾大千软装设计师在不同空间的色彩定位上进行精心雕琢，客厅中深邃的蓝色，第一眼便震住一切的夺魄气势，彰显着主人的威严权势和身份地位。同时，蓝色的宁静，也将繁华与华丽背后更深层次的温暖和品质，以及无穷无尽的遥远梦幻，智慧地带入眼前的最私密、最温馨又最完美的家世界。

元禾大千软装设计师精选装饰品与艺术品，与整体空间的气质相呼应。精致唯美又不失自然雅趣的顶级中式图案缩影，传统丝质面料，意蕴绵长的东方饰物，还原生活本真追求，彰显一个时代的辉煌，一个圈层的荣耀。水晶、黄铜、银器，精美花艺等材质使其气质璀璨缤纷，富丽堂皇，加倍凸显华贵的工艺品质，也充分体现了精湛的欧洲工艺水准和高度提炼的装饰细节。

餐厅：

　　典雅大方的陈设艺术品，或是细腻柔美的线条，无论是材质本身的特点，还是物品的多元化组合，每一处蜿蜒都呈现出大气之美，呈现出耐人寻味的意境和文化传承。放眼望去整个餐厅空间无论是家具还是窗帘、地毯、饰品之间的色彩搭配，都充分体现浓郁的新古典魅力，让空间中的色调更加饱满、生动、统一。

书房：

偌大的书房空间呈圆形勾勒，宽敞大气。规整的天花顶面处理，扇形排开，配上奢华的水晶吊灯，正式感加倍。同时配以嵌入式墙面实木书柜，实用美观，序列感强烈。壁炉的设计，突出了风格的所在。上方古典欧式造型的人物图像，更显气质。

酒窖：

　　沉稳有格调是设计师给予酒窖的定位，因此设计师特别注意软、硬装的搭配，整体以突出硬朗的硬装材质和舒适的软装质感为协调。色彩上偏明度低的黑色和深咖色，如壁纸、地毯、沙发等，而跳跃的黄色、蓝色与金色又巧妙的点缀了空间的沉闷，打造出一处既私密又富有情调的休闲空间。

SPACE PLANNING ｜ 空间规划

　　整个别墅空间多达1700平方米，整体规划详尽合理。挑高的客厅贵气开阔，多餐厅的设置方便贴切，同时中西融合的设计手法更显老练大气。动静分区，公私分明。各个空间都极好的利用了自然采光的优势，处处都能感受到室外的春意盎然。

主卧：

精致的工匠艺术是卧室吊顶给我们的最大感受，一笔一线都尽显优雅与贵气。排排相连的花卉图案似古时的钱币，加上镀金的效果，更显逼真。其中三层圆筒水晶吊灯明亮净透，空间感更加豪华舒适。而真丝绸缎的材质配上精美的刺绣，艺术品味不言而喻。

儿童房：

粉嫩的色彩配上艺术的金色线条让空间氛围更显气质。不规则镜框、床头背景墙以及台角等，都很好的衬托了屋主的审美。而各种鲜花、动物、雕像元素的加入，则展现了屋主的生活乐趣。